THE PHYSICS OF SPIRITUAL MOMENTUM
ISBN 13: 978-1-946467-03-4
ISBN 10: 1-946467-03-0
Copyright © 2019 by JACK SHOUP
4514 Briar Hill Rd.
Lexington, KY 40516

Published by:
Piercing Light Publishing
4514 Briar Hill Rd.
Lexington, KY 40516
502-570-9343
Printed in the United States of America.
All rights reserved under International Copyright Law. Contents and/or cover may not be reproduced in whole or in part in any form without the express written consent of the Publisher.

All Scripture quotations are taken from the *King James Version* of the Bible

Cover design by Troy D. Ledford www.bluewhaleinnovation.com.

THE PHYSICS OF SPIRITUAL MOMENTUM

by
Jack Shoup

TABLE OF CONTENTS

Page

The Concept	*1*
The Physics of Faith	*5*
Spiritual Momentum	*23*
The Works of God	*35*
Divine Energy	*43*
The Power of God	*53*
Conclusive Results	*63*
About the Author	*67*

CHAPTER 1

THE CONCEPT

Many, if not most people have a very limited understanding of the Kingdom of God. Even Christians are often mistaken to believe that God's kingdom only represents the place where the righteous go after they die. However, Scripture makes it clear that access is granted to this kingdom once an individual is born-again – *not when they die*. The blessings of healing, spiritual authority, prosperity, and victory belong to us today. However, it is not sufficient to know that God has made available to us access to His divine benefits. We must be able to apply Kingdom principles that allow us to manifest these blessings.

The average Christian has little comprehension of the specific steps necessary which empower their prayers to be answered. They go through their life with a mental image that Heaven's ways are far too high for us to understand. Therefore, they live with the mindset that having their prayers answered is somewhat like flipping a coin. They believe that sometimes God answers our requests, and other times He does not – depending on how the winds from Heaven are currently blowing. However, little could be further from the truth.

Bible believers understand that our natural world and the universe that we currently dwell in were created by God out of the spirit realm. Know also, that **everything in this natural universe functions by established principles that we call the *Laws of Physics***. These physics laws include principles that govern gravity, thermodynamics, electricity, chemistry, motion, and so on. Everything that takes place within this natural realm follows these set laws. The laws of physics are so reliable, repeatable, and precise that we can use them to send men to the moon and bring them back safely. We can build planes that don't suddenly drop out of the sky for no apparent reason. The laws of physics for all general purposes are unbreakable.

However, this natural predictable world was birthed out of the spirit realm. So, if our natural universe is firmly upheld by unbreakable laws, it would be ludicrous to believe that the laws of the spirit are not equally reliable. In fact, the Kingdom of God is more immovable and predictable than this universe in which we currently abide. Jesus declared that the Heavenly realm will outlast the world as we know it today.

Mt 24:35 - "Heaven and earth shall pass away, but my words shall not pass away."

The Kingdom of God functions by spiritual laws in the same way that this universe operates by natural laws. These spiritual laws, if understood and activated, will produce supernatural results as surely as using natural laws to produce anything on earth. However, if we do not know or understand these spiritual laws, we cannot reliably manifest promised Kingdom blessings. For the first 5900 years of man's existence on earth, we did not know how to build or power an airplane. However, once we came to understand the principles of *lift* and *thrust*, air travel became commonplace and opened the world to great opportunity.

In these last days, God is unveiling to His church key spiritual laws or principles that will allow us to transform adverse conditions all around us and bring much of Heaven's blessings to earth. Once we comprehend these spiritual principles, our prayers will be infused with power and produce *guaranteed results*.

The spiritual laws of faith, love, sowing and reaping, forgiveness, and numerous others all allow us to live victorious lives once we activate them.

This book will not be a comprehensive look at each of these spiritual laws. Instead, it will take a closer look at a few natural physics principles to examine them for potential spiritual significance. It occurred to me several years ago to look at some of the natural laws that govern our world to see if they had any parallels to the laws of the spirit. The information and formulas presented in this book may not be exact representations of the spiritual laws to which they are compared. However, they will provide us with an insight into how some into God's kingdom principles function and can be utilized.

In this book, we will focus primarily on laws of physics that govern motion and their possible spiritual counterparts.

Note: Do not be intimidated by the topic of physics and the threat of looking at the different scientific formulas to be covered. If anyone has difficulty understanding the mathematics used to derive some of the formulas covered in this book, just feel free to skip that part and read the sections covering the spiritual significance of those formulas

Most of the topics covered in this book have been simplified to make them as easy to understand as possible.

CHAPTER 2

THE PHYSICS OF FAITH

The First Law of Motion

Sir Isaac Newton postulated three primary laws of motion. His first law of motion describes what we refer to as *inertia*. This natural law tells us that any object *at rest* or *in motion* will remain in that state of rest or motion unless it experiences a *force* applied to it. For example, a car going down the highway will continue to roll even if we let off of the gas pedal. Inertia will keep that car in motion until wind resistance, road friction, or brakes bring it to a stop. In other words, regarding motion, nothing changes unless forces are applied to produce a shift in status. If no forces are applied, everything stays the same. Because of inertia, not only is a force required to place an object in motion, but an opposing force is also required to stop an object that is already in motion.

This same law of inertia is also true regarding changes to be accomplished on Earth supernaturally. No supernatural change on earth proceeds from God's kingdom unless force is applied to initiate that change. However, the force that we apply to enact spiritual change is not produced by the brute strength of human effort. The force that initiates supernatural spiritual shifts from God's kingdom is *Faith*. In fact, faith becomes a primary determinant of anything that takes place in God's kingdom. No Heavenly initiated change is released to earth without faith producing the shift. We see this potential of faith displayed in several scriptures.

Mk 11:22-24 - "And Jesus answering saith unto them, Have faith in God. For verily I say unto you, That whosoever shall say unto this mountain, Be thou removed, and be thou cast into the sea; and shall not doubt in his heart, but shall believe that those things which he saith shall come to pass; he shall have whatsoever he saith. Therefore I say unto you, What things soever ye desire, when ye pray, believe that ye receive them, and ye shall have them."

We have been given authority to speak words of faith and watch mountains move from before us. God has established spiritual laws regarding faith which allow us to access Heavenly power. This Heavenly power subsequently produces desired change in our lives. In Matthew, Jesus declares:

Mt 11:12 - "And from the days of John the Baptist until now the kingdom of heaven suffereth violence, and the violent take it by ***force****."*

The supernatural forces that Christians are to use to produce changes in this natural world are only accessed through faith. Faith is the key to accessing the forces of the Kingdom of God. Faith is not the force of itself. However, faith is the trigger that releases God's power to produce desired changes around us.

To better understand the role of faith, think of it as the accelerator pedal to your car. The pedal has no inherent power of its own but when depressed, it allows the engine to release the power built into it. Our faith, when activated, releases the power of God's Kingdom to accomplish the supernatural. As well, the harder that we press on the pedal, the more power the engine produces.

Imagine this pressure on the pedal represents operating in ever-increasing levels of faith. As well, the longer that we keep the accelerator pedal depressed, the further and faster the car will go. Let this represent the continued or prolonged application of faith.

In fact, without faith, we cannot even begin to fulfill our genuine calling from God.

Heb 11:6 - "But without faith it is impossible to please him: for he that cometh to God must believe that he is, and that he is a rewarder of them that diligently seek him."

Newton's first law of motion regarding inertia can be reworded to become a Kingdom law.

**THE FIRST LAW OF KINGDOM TRANSFORMATION -
No situation or circumstance on earth will be altered by the supernatural power of God until someone releases** *force* **through the operation of** *faith* **in God's Word.**

Once the forces activated by faith are released into our situations, anything is possible.

> *Mt 17:20 - "And Jesus said unto them, Because of your unbelief: for verily I say unto you,* ***If ye have faith as a grain of mustard seed****, ye shall say unto this mountain, Remove hence to yonder place; and it shall remove; and* ***nothing shall be impossible unto you.***"

Through faith, we can access every blessing and promise of God. However, until we develop and utilize our faith, no supernatural change may take place. The spiritual law of inertia demands that faith be released to produce desired changes. For the remainder of this book, when we utilize the letter 'F' in any 'spiritual' equation, we are referring to the **forces from heaven that are activated by faith**.

The Second Law of Motion

Newton's second law of motion deals with the *measurable effect* of applying a force to an *object* – the *object* being something that has *mass*. Newton declared that when a force pushes on a mass of any size, unless countered by an opposing force, that mass will accelerate in a predictable fashion. For example, if we place a bowling ball on a level surface and push it, that ball will move in proportion to how much force we apply.

A Force Applied Causes the Bowling Ball to Move

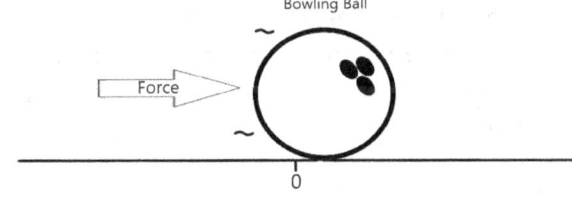

We thus imparted a force to a mass and it accelerated accordingly. In fact, anytime any object is put into motion, or even if that motion changes in any way, it is because a force has been applied to it. This second law of motion is precise and repeatable and can, therefore, be written in the form of a formula.

To write laws in formula form, we assign different letters to equate to each component of our equation. In this case, we want to examine the result of applying a **force (F)** to a **mass (m)** which produces a subsequent **acceleration (a)**. In formula form, Newtons second law of motion is stated as:

$$F = ma \quad \text{or} \quad F = m \text{ times } a$$

What this formula tells us is that the harder that we push on an object, the faster it will accelerate in the direction[1] of the force applied. It also lets us know that the larger the mass of an object is, the more force it will require to move it at an equivalent rate. It takes more force to equally accelerate a car than a bicycle.

So, how can this be carried over and applied to our examination of the laws of the Kingdom of God? Recall that the *forces* that we apply to produce supernatural change are those activated by *faith*. Now bring to mind what we just read from Mark 11:23 - that faith holds the potential to move mountains. *Mountains* are very heavy and represent the *mass* that we want to move.

However, when Jesus was speaking of moving mountains, He wasn't predominantly focused on moving actual earthen mountains. Those mountains that we are empowered to move by faith represent whatever problems stand in our way - regardless of their size. So, faith carries the potential to deliver us from any challenge that we may face in life. As we apply faith to the adverse circumstances that arise against us, those mountains must accelerate out of our path.

So then, we want to let the *heavenly forces that faith activates* be **F**, and the size of the *mountain* or *challenge* be **m**, and the *rate at which our problems are being moved out of the way* be **a**. As we plug these into Newton's second law of motion we get:

---

1 Vector analysis will not be covered in this book.

F = ma or
Faith = mountain size x rate the mountain accelerates out of the way

As we examine this spiritual law and equate it to our prior examination of the laws of motion, we can deduce some important truths.

First - The level of faith that we apply will determine how fast our challenges may be resolved supernaturally. It will take longer to see resolutions to our problems with low faith levels than with High.

Second - Bigger mountains (challenges) take more faith to move in the same amount of time as smaller ones.

Third - If no faith is applied, no supernatural change is going to take place.

As Christians, this should point out the importance of building up our faith. Although we are all imparted a measure of faith when we are born again, it is our responsibility to build this small measure into *great faith*. Without advancing in our faith development, we will be unable to address any significant challenges that lie in our future. Just as a baby is to grow into adulthood, we are assigned of God to develop our faith in His Word once saved.

The primary way that a Christian is to develop their faith is through meditation of God's Word.

Rom 10:17 - "So then faith cometh by hearing, and hearing by the word of God."

By spending time reading and meditating Scriptures, a miraculous phenomenon takes place. The word becomes engrafted to our souls. The scriptures, that before were merely words on a page, become real to us. This newfound belief in God's Word reshapes our expectation of what is possible for us to accomplish.

Through faith, amazingly, the impossible becomes doable. Through this newly developed faith, we expect the promises of God that we voice forth to come to pass.

For the Christian who wants to fulfill the call of God on his life, building faith becomes a necessity. Large mountains will certainly rise against us. Large faith will be required to move them out of the way in a reasonable period of time.

Graphing the Fight of Faith

One of God's most important promises to the body of Christ is that, through Jesus' sacrifice, we have access to supernatural healing power.

> *1Pet 2:24* - *"Who his own self bare our sins in his own body on the tree, that we, being dead to sins, should live unto righteousness:* ***by whose stripes ye were healed.****"*

God's healing power is being poured out now as never before. Multitudes are being physically healed from every type of infirmity that exists. However, many misunderstand how divine healing manifests when people are prayed for. There are two primary ways that supernatural healing takes place in the church and confusion between the two causes some to miss out on this blessing.

These two paths to healing are:

1) Through the **gifts of healing** and
2) Through **faith in God's Word**.

The *gifts of healing* are mentioned as one of the nine spiritual gifts in First Corinthians.

> *1Cor 12:7-11* - *"But the manifestation of the Spirit is given to every man to profit withal. For to one is given by the Spirit the word of wisdom; to another the word of knowledge by the same Spirit; To another faith by the same Spirit; to another the **gifts of healing** by the same Spirit; To another the working of miracles; to another prophecy; to another discerning of spirits; to another divers kinds of tongues; to another the interpretation of tongues:*

But all these worketh that one and the selfsame Spirit, dividing to every man severally as he will."

The gifts of healing are exactly that - they are gifts. The gifts of healing may manifest regardless of the faith of the one being prayed for. Instead, the gifts of healing function based primarily upon the faith of the one doing the praying. In most cases, the results are immediate often to the surprise of the recipient. What a wonderful blessing – to come into a prayer line sick, in pain, or with some form of disability, and to leave symptom free.

However, there are a few problems that exist for the believer who desires to be healed by the gifts. First, not every minister flows in the gifts of healing. It may be difficult to locate someone to pray for them. Just because a gift exists, does not mean that there is an available *gift-carrier.*

Also, for those who attend a major crusade where the power of God is moving to heal, there is no guarantee that they might be called forth to receive of the gift. Many healing crusades pack thousands into a building only to see a handful given access to the platform. Many more often return home with no change in their health status. I am certainly not trying to demean healing crusades but merely want to point out that they provide no assured path to recovery.

There is, however, a reliable way to receive our healing. The sure-fire method to be made whole is via faith in God's Word which promises us healing. By developing our faith to be healed based on God's promises in the Word, we are provided with a guaranteed path to the restoration of health. Once we learn to be healed by believing and speaking forth what God has declared in His Word, we won't need to track down the nearest healing crusade hoping that our number is called. Faith provides an avenue to God's healing power that produces results every time.

Yet, there are a few hurdles that exist when we endeavor to be healed by faith:

> **First -** We have to develop our faith for healing by spending time meditating, from Scripture, God's promises to heal us.

Second - Healings that take place by faith are not always instantaneous – they usually involve a time factor.

In fact, it is not unusual for symptoms to grow worse after we have decreed our personal healing. Healings that are received by faith often require us to diligently stand on God's Word until we see the manifestation. Such healing results require us to apply the force of faith until symptoms are reversed.

To demonstrate this, imagine a locked car that has slipped out of gear and begins rolling down your gently sloped driveway.

Car Begins to Roll Down Driveway

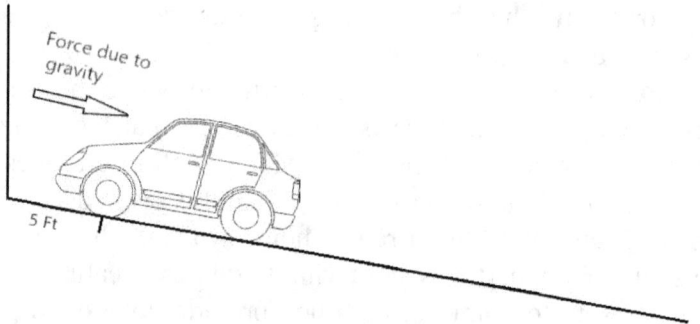

To stop the car, we must get behind it and apply a force greater than the force of gravity pulling it down the drive. Assume that we do this after it has already rolled 5 feet. Once we apply the force, the car does not stop immediately. Because of the law of inertia. it takes time to reverse the roll and bring the car to a stop. We can see this effect in the following image:

Man Stops Car from Rolling Backwards

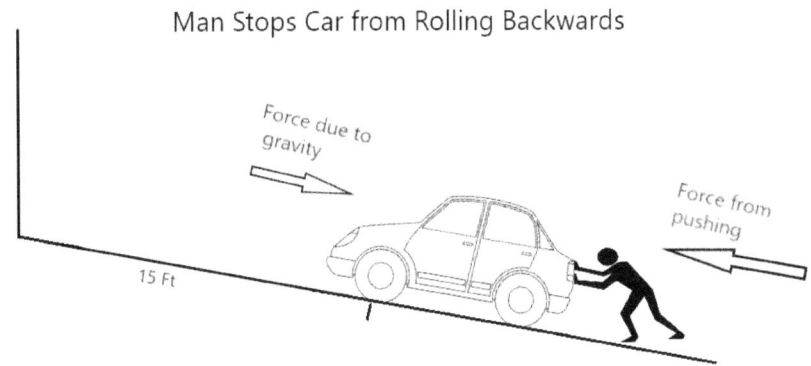

Notice that the car traveled an additional 10 feet after we began pushing before it was halted.

Recognize as well, the harder that we can push (per Newton's second law) the faster the car will come to a halt. If a stronger man applies force to the car, it will stop more quickly.

Stronger Man Stops Car in Shorter Distance

In this case, the stronger man was able to halt the car in half the additional traveled distance.

Now imagine that the car represents physical symptoms of sickness. As we apply heavenly force through faith to reverse the symptoms, they may not disappear immediately. Faith must be maintained over time until the sickness is totally defeated.

Looking at this graphically:

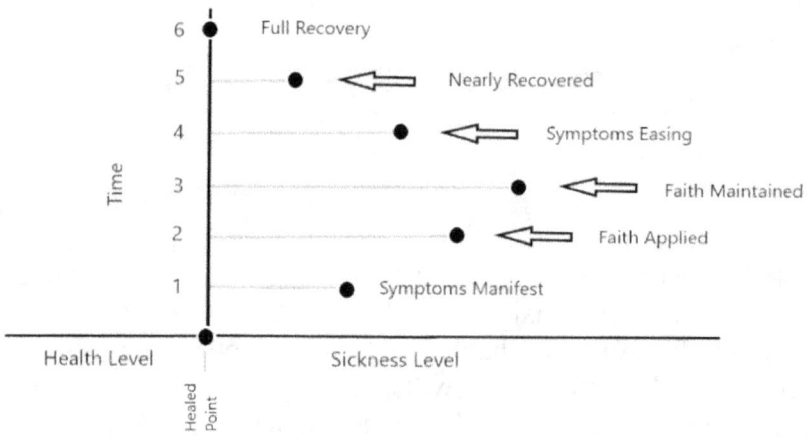

Faith Application Recovery Chart

The *healed point* on this chart indicates the health level at which we are symptom-free. Any condition to the right of the healed point indicates symptoms of sickness in our bodies. The further we are to the right of the healed point indicates more severe symptoms. The vertical bar labeled *Time* can be said to represent progressive time segments (hours, days, weeks, etc.) we are confronting sickness.

The example presented with this chart shows us waking up on period 1 with symptoms of sickness present. Since it often takes us a bit of time to recognize that we may be ill, in this case, we do not begin to apply our faith to be healed until period 2. However, as with the car, the symptoms carry some inertia of their own. Even though we have released our faith to be healed, it is not uncommon to wake up feeling worse in period 3.

This is the point many Christians attempting to use their faith for healing quit. Since they feel worse after having released their faith, they become convinced that it's not working. However, because they do not comprehend the spiritual physics in operation, they quit too soon.

Notice, that if our faith is kept active, in period 4 the symptoms are reversing. Then by period 6, we experience a full recovery.

Also, just as the stronger man can stop the rolling car in a shorter distance, stronger faith allows us to reverse symptoms in our body more rapidly as well. We see this in the chart below.

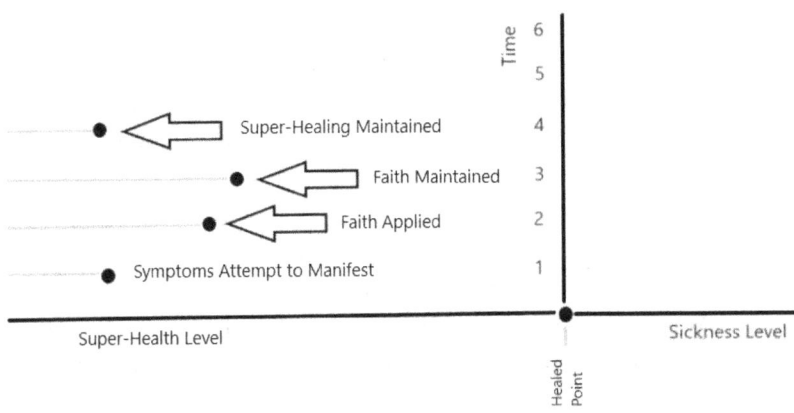

Super-Health Application Chart

Recognize in this example that healing is accessed in 4 periods rather than 6. Notice as well, with greater faith, not only are all symptoms driven out of the body in a shorter period of time, the symptoms also do not grow as severe. This brings us to a second law of Kingdom operation which parallels the laws of physics.

**A SECOND LAW OF KINGDOM TRANSFORMATION –
As our faith increases, we can move larger and larger mountains in shorter amounts of time.**

From this example, we should be able to see the need to grow our faith even when we are not experiencing physical attacks. Because *great faith* can fight back symptoms much more effectively than *small faith*, it will be of great benefit to build our faith even before we are sick. Not only that, but developing our faith while we are healed is much easier than when we are ill. The symptoms of sickness make it difficult to meditate on anything other than the pain and discomfort.

So many Christians fall into the trap of waiting until they are sick before they will build their faith for healing. It is so much easier to block the wheels of a car before it begins rolling than to try to stop it once it is in motion. The same holds for manifesting divine healing. This will be discussed in more detail in our chapter on *Spiritual Momentum*.

Other Christians miss out on the opportunity to develop and exercise their faith for healing by always running to the medicine cabinet at the first sign of discomfort. Make no mistake, I am not against doctors or medicine. However, it is beneficial to learn to build and utilize our faith against small health issues rather than wait until something life-threatening occurs. Before David killed Goliath, he first took on the lion and the bear. If we can learn to obtain victory over headaches and stomach aches, we will be far better prepared to tackle more serious symptoms in the future should they occur.

There is an additional benefit of using our faith for small challenges rather than waiting to be overwhelmed by a large one. Small faith victories produce personal *experience* which builds our expectation of victory, even as bigger mountains arise before us.

Another problem of waiting until the last minute to build faith occurs when people experience potential terminal illnesses that the medical community cannot mend. Remember that, due to inertia, symptoms often grow worse once we begin to apply our faith to reverse them. If a person, by faith, tries to reverse the symptoms of something that threatens their life, often that person may run out of time to fully enact their recovery before the disease claims them. We see this potential shown in the next chart.

Underdeveloped Faith Applied to a Critical Condition

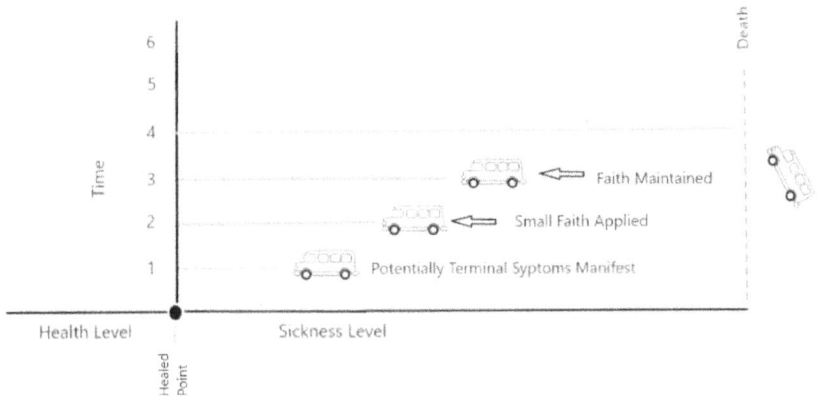

In this example, life-threatening symptoms hit the Christian, however, this believer possesses only small faith. The symptoms are of such a high level that his faith cannot reverse the backward momentum before his situation becomes terminal.

How much easier it is to build great faith before the adverse effects of disease takes us over a cliff. We should not wait until it is too late to build our faith for healing, however, this is somewhat true for all of God's promises. Why wait for debt to overflow our lives before we take the time to build our faith for God to financially bless us. Why wait until our family members are on their death bed before we claim them for the Kingdom of God. We can apply the spiritual principles of faith in advance of any desperate situations which may arise in our lives.

Scripture alludes to our ability to be healed or delivered in an accelerated fashion.

*Psa 31:2 – "Bow down thine ear to me; deliver me **speedily**: be thou my strong rock, for an house of defence to save me."*

*Isa 58:8 - "Then shall thy light break forth as the morning, and thine health shall spring forth **speedily**: and thy righteousness shall go before thee; the glory of the Lord shall be thy rereward."*

Offensive Faith

When physical symptoms hit our bodies, those symptoms may immediately place us in a defensive position regarding our health. However, by building up and using our faith in advance of sickness, we can take up an offensive position. We can repel symptoms before they may make us ill.

Not only has God given us promises of divine healing but also of *divine health*. Divine health declares that we never fall prey to sickness. How much better is this than being sick and then trying to apply our faith to recover. This opportunity is clearly spelled out in Scripture.

> Ps 91:10 - *"There shall no evil befall thee, neither shall any plague come nigh thy dwelling."*

Symptoms may try to impact us when we are in divine health but they cannot remain on our bodies. They fall off as if we were non-stick cookware. We may refer to this divine health position as being *super-healed*. We can see this graphically:

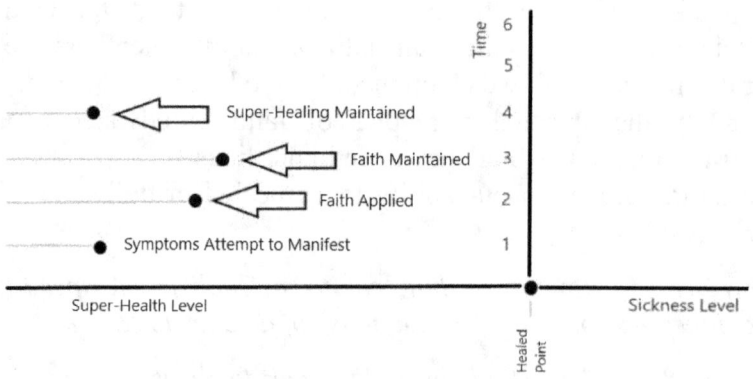

In this example, the believer has built his faith to a very high level for divine healing. Even though the symptoms try to hit, they cannot drive the faith-filled individual behind the healed point.

His health remains intact. The subsequent faith that he releases continues to work to keep him in the super-healed position.

Our faith in divine health can place us in a position of *super-healing*. Just as sickness knocks the average person below the healed point, super-healing positions us above the influence of sickness. How wonderful to be established in such a blessed position. Through the application of faith, we can live a life where we never have to be sick again.

This may be more easily understood if we apply this concept to our finances. It is far preferable to be debt-free with a positive cash flow and an adequate savings account than to be living paycheck to paycheck. When living week to week, it only takes one mishap or breakdown to throw one further into debt. However, when we have built up a savings account, we can take the unexpected financial hits with little impact on our lifestyle. Divine health by faith operates similarly.

Those who *live by faith* don't wait until things become desperate to apply their faith. Applied faith has the potential to keep us empowered to overcome every assault of the enemy - including sickness.

> *Eph 6:16 - "Above all, taking the shield of faith, wherewith ye shall be able to quench all the fiery darts of the wicked."*

End-Time Acceleration

By further examining our formula for faith, we discover how God must increase the faith of His church in these end-times. We have already found that if we have great faith, then we can move larger mountains. However, great faith also enables us to accelerate the rate at which the mountains are moved. As we have seen, a high faith level can reverse symptoms more rapidly than a low faith level. This obvious truth becomes a key for God to finish His work to perfect His Bride in these last days. God, even now, is pouring out increased revelation of His Word as never before. As mentioned, faith comes by hearing and hearing the revelation of the Word. For those with a heart to pursue God fully, this increased revelation is causing the faith of His End-Time Church to rise to unprecedented levels. This elevated faith is allowing the Church to complete the works of God in an accelerated fashion. In fact, God must increase the development of His Church in an accelerated fashion to equip them to complete their assignment to fill the World with His glory.

> *Hab 2:14 - "For the earth shall be filled with the knowledge of the glory of the Lord, as the waters cover the sea."*

The Word lets us know that God has a two-fold assignment for His Church to fulfill in these last days.

One – He expects His Church to dominate and crush the powers of darkness before the Rapture occurs.

Two – God has assigned the Church the responsibility to bring in an unprecedented harvest of souls.

Both of these assignments placed upon the Church will demand great manifestation of the miraculous mountain-moving power of God. As well, if the Rapture takes place as soon as many believe, then time is running out to complete these tasks.

The church obviously has a long way to go in their spiritual development and subsequent manifestation of power but only a short time remaining to accomplish it.

Therefore, God is accelerating all things pertaining to the development of His Church.

Paul alludes to this acceleration of the development of the church in the book of Romans.

> *Rom 9:28 - For he will finish the work, and cut it short in righteousness: because a short work will the Lord make upon the earth.*

By pouring out the revelation of the Word as never before, the faith level of the Church will rise to mountain-moving capacity.

As the faith level of the Church soars, their ability to dominate the forces of darkness and bring in God's harvest will be accelerated.

What a joy it will be to be used by God in such a previously unparalleled fashion. As verified by our formula, the forces initiated by faith become key to accelerating the prophesied role God's Church is to fulfill.

Throughout this book, we will witness the role that faith plays in every area of spiritual application that we endeavor.

CHAPTER 3

SPIRITUAL MOMENTUM

In this chapter, we are going to delve a little deeper into our study of motion and investigate some potential spiritual implications that exist. Recall our formula of Newton's Second Law:

$$F = ma$$

or

Faith = mountain size x rate the mountain accelerates out of the way

A new aspect of our study that we want to bring into focus is the quality of motion referred to as velocity (v). Velocity represents the speed an object is moving at any point in time. While acceleration indicates that an object is speeding up or slowing down, velocity is the instantaneous measure of that speed. If a car is accelerating at 10 miles per hour for every second that we press on the gas pedal, then its acceleration is 10 mph per second or 10 mph/sec. If we hold down the gas pedal of this car beginning from a stop, after 5 seconds, the car will be traveling at a velocity of 50 mph. The formula for velocity as it relates to acceleration is:

velocity = acceleration x time

or

$$v = at$$

The implications of this are obvious. The longer that we accelerate an object, the faster it will be moving. Regarding our mountain that we are moving by faith, the longer we apply our faith with consistency, the faster it will disappear from before us.

Now for some fun with algebra! If we take this formula, we can algebraically divide both sides by **t**.

$$v = at$$

dividing both sides by **t** we get

$$v/t = at/t$$

by canceling out the **t**'s on one side we get

$$v/t = a$$

This simply means that if we wanted to find the average acceleration of an item in motion, we divide its final velocity by the time that it took for us to reach that speed from a dead stop. If a car can accelerate from 0 to 60 mph in 6 seconds, then it is accelerating at 60/6 = 10 mph/sec.

Now let's perform a similar operation on our former formula for acceleration.

$$F = ma$$

dividing both sides by **m** we get

$$F/m = ma/m$$

again, now the **m**'s cancel out and we get

$$F/m = a$$

This formula tells us that our acceleration is equal to the magnitude of the force that we apply divided by the mass of the object that we are moving. To accelerate a large object requires more force than to accelerate a smaller one.

Now more fun with algebra. We can now put these two formulas together.

Since $a = F/m$ and also $a = v/t$

therefore

$$F/m = v/t$$

A primary rule of algebra allows us to multiply both sides of an equality equation by the same amount and still have the equality hold true.

So by multiplying both sides by **mt** we get

$$Fmt/m = vmt/t$$

simplifying algebraically we get

$$Ft = vm$$

or **Force x time = velocity x mass**

This is the classical formula for the relationship between *impulse* and *momentum*; where **Ft = impulse** – the application of a force over time, and **vm = momentum** – the product of the velocity of an object times its mass.

What is impulse?

Impulse represents the application of a force over a period of time. In our previous examination of accelerating a car, impulse represents how long we hold down the gas pedal. The longer that we hold down the pedal, the more impulse we have imparted to the car, and the faster the car will be going. Because momentum is related to the velocity of the car, impulse therefore alters that momentum.

What is Momentum?

In our previous chapter, we examined Newton's first law of motion which deals with *inertia*. In this law, he declared that an object at rest or in motion will remain at rest or in motion unless acted on by a force. *Momentum* can be thought of as a measure of that inertial capacity. Momentum quantifies the difficulty in altering the movement of an object at rest or in motion.

The implications of momentum should be somewhat obvious to us. Momentum makes it much more difficult to stop a 20,000 ton truck traveling 30 mph than a 40-pound bicycle traveling the same speed. The truck will have 1000 times the momentum. Therefore, momentum is often considered to be the *quantity of motion* of an object.

The more that an object weighs combined with the velocity that it is traveling both establish the momentum of that object. An object with high momentum is able to plow through obstacles that objects with low momentum cannot. Momentum allows a 12-pound bowling ball thrown at 20 mph to knock over some pins while a 40-pound cannonball projected at 500 mph will knock down buildings. Objects with high momentum are much more difficult to stop.

Spiritual Momentum

Just as objects may be acted on by a force over a period of time to produce high momentum, we can apply the forces activated by faith to produce spiritual momentum. If we allow our formula for momentum to represent spiritual attributes, it may provide us with additional comprehension of the importance of building and releasing our faith. Recall that our formula for momentum is:

$$Ft = vm$$

If we again allow **F** to equal the *supernatural forces accessed by faith* and **t** to equal the *time* our faith is applied, then **vm** equates to our level of *spiritual momentum*. Just as an object with high natural momentum may bowl over or shatter objects that stand in their path, **Christians with high spiritual momentum may plow through every form of sickness, pain, or oppression that tries to oppose them.** Conversely, those with low spiritual momentum may be stopped in their tracks at the first sign of pain or sickness.

A THIRD LAW OF KINGDOM TRANSFORMATION - Spiritual momentum enables the believer to push back and overcome every attack of the enemy.

Because natural momentum is built up by applying a force over an extended time, spiritual momentum is increased by keeping our faith activated for a promise of God over a period of time. An important area to build up our spiritual momentum is again for divine health. Rather than wait for illness to strike our bodies, **we can regularly confess Scriptures that promise us healing**

before we are sick!

Of those Christians who have an awareness that God will heal their infirmities by faith, many often do not begin applying their faith until adverse circumstances arise. In such cases, they possess little spiritual momentum and will almost always experience a setback from the physical symptoms. However, a believer that will spend time daily confessing the healing promises of God over their life may build up positive spiritual momentum for their health. This momentum enables them to plow right through any symptoms that may arise. We looked at this in the former chapter in our discussion of *super-healing*.

Physical symptoms and adverse situations carry their own form of momentum that opposes the promises and power of God. Once physical symptoms are able to enter our bodies, they can tend to overwhelm us mentally. Thus, if the enemy can get us to entertain fear of symptoms or doubt in the promises of God, it acts to curtail our spiritual momentum.

A wise Christian will spend time confessing the Word of God over their life such that adverse circumstances do not even slow them down. Through this application of developing positive spiritual momentum, the believer can step into a lifestyle of living in *divine health*. This is far superior to attempting to access divine healing once we have succumbed to symptoms and they have imparted their own negative momentum. **Divine healing** allows us to receive healing from whatsoever type of sickness or disease that may try to attack us. However, through **divine health**, we can develop a spiritual momentum whereby we never have to be sick another day of our lives.

To demonstrate the significance of possessing spiritual momentum, let's look at some natural examples of momentum in action. Imagine a bowling ball sitting still on a perfectly level surface and that a ping pong ball is projected at this stationary ball.

Ping Pong Ball Launched at Stationary Bowling Ball

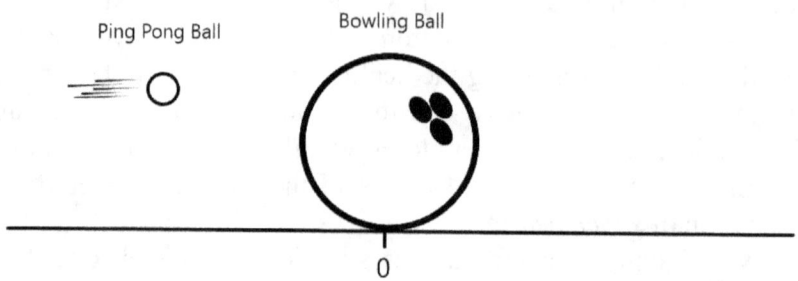

Now imagine that the ping pong ball strikes the bowling ball and bounces off of it.

Ping Pong Ball Causes Bowling Ball to Move Backwards

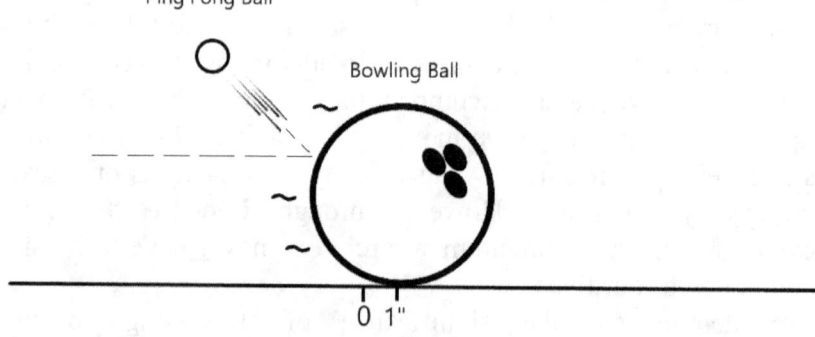

Even though the bowling ball is much larger, because the ping pong ball carries momentum, it applies a force to the bowling ball and transfers a portion of its momentum to the larger ball. According to Newton's second law of motion, **F = ma.** Because a force is applied to the bowling ball, it will begin to move.

Now let's relate this example to our ability to supernaturally combat illness. If we are releasing no faith to oppose possible attacks of sickness prior to the appearance of symptoms, then we will possess no spiritual momentum against such attacks once they occur. From a spiritual momentum standpoint, we are standing still. Thus, whatever physical symptoms come against us will almost certainly push us back into being ill. This is the state of most Christians who may know about faith but take no steps to build or apply their faith until after they are sick.

Assume however, that we take that same bowling ball and roll it at a reasonable speed toward the oncoming ping pong ball.

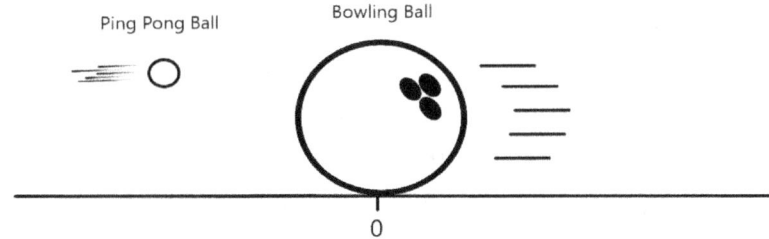
Ping Pong Ball Launched at Bowling Ball in Motion

Once in motion, you can launch numerous ping pong balls against it and barely slow it down.

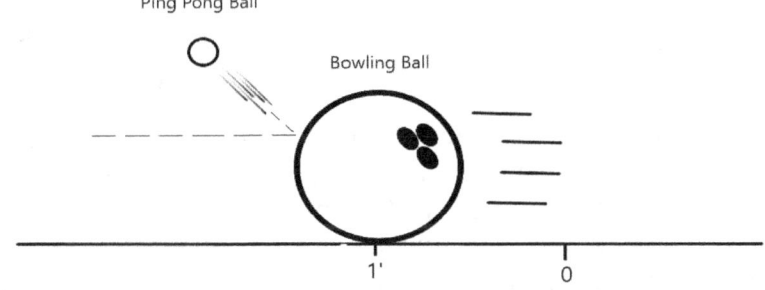
Ping Pong Ball Has Little Impact on Bowling Ball In Motion

The momentum of the bowling ball is more than sufficient to repel a ping pong ball unless it is launched at an extremely high speed.

Spiritual momentum works in the same fashion. Once we have set our faith in motion for divine health (declaring that sickness and disease are far from us), most symptoms will bounce off of us like shooting bb's against an army tank. Spiritual momentum allows us to truly declare:

Isa 54:17 - "No weapon formed against us shall prosper..."

Another great application of this principle is that we can keep building and releasing our faith until we are operating in *Great Faith*! Why stop at minimal levels of spiritual momentum when we can build up so much spiritual inertia that no sickness can even come near us. Thus is fulfilled the amazing promise from Psalm 91.

Psa 91:10 - "There shall no evil befall thee, neither shall any plague come nigh thy dwelling."

The word *plague* in this verse is the Hebrew word *nega* which means injury, sickness, or wound. Spiritual momentum can bring us to the point that every physical attack of the enemy is repelled.

In addition to this, we can build spiritual momentum for ALL of the promises of God. In this fashion, not only do physical attacks bounce off of us, but emotional onslaughts are neutralized as well. What an awesome God we serve!

We can see the tie between spiritual momentum and the ability to receive the promises of God reflected in the book of Hebrews.

*Heb 6:12 - "That ye be not slothful, but followers of them who through **faith and patience inherit the promises**."*

The word patience ties our activities of faith to the passage of time. Therefore, this powerful verse of scripture lets us know that we can receive the promises of God by applying our faith over a period of time. As we have found, this faith applied over time can be expressed mathematically as **Ft**. **Ft** has already been defined as *impulse*. As we apply our faith over a period of time, we are imparting *spiritual impulse*. Since we have also learned that **Ft = vm**, or *momentum*, then **it is spiritual impulse that produces spiritual momentum that allows us to receive the promises of God.** Surely we are beginning to see the importance of developing this unstoppable spiritual quality.

A FOURTH LAW OF KINGDOM TRANSFORMATION – As we spend extended time building our faith and confessing God's Word, we build spiritual momentum.

Patience allowed Abraham to receive the promise of a son:

*Heb 6:15 - "And so, after he had **patiently** endured, he obtained the promise."*

Patience becomes a key to have the seed of God's Word fully produce in our lives:

*Luke 8:15 - "But that (seed) on the good ground are they, which in an honest good heart, having heard the word, keep it, and bring forth fruit with **patience**."*

In this fashion, patience enables us to receive the promises of God ourselves:

*Heb 10:36 - "For ye have need of **patience**, that, after ye have done the will of God, ye might receive the promise."*

From this, we can see just how vital it is to keep our faith intact and operational. It is not enough merely to voice a promise of God. We must continually keep a positive attitude and a consistent confession regarding that which we have declared is ours by faith.

Building Spiritual Momentum

As we have discovered, the relationship between impulse and momentum is expressed as:

$$Ft = vm$$

From this, we have ascertained that momentum is altered by

applying a force to an object over a period of time. Once an object has been placed in motion, the longer that a force is applied to it increases its momentum proportionally.

The same truth holds for releasing our faith in the promises of God. The more faith that we can consistently release, the more spiritual momentum that we subsequently build. **Conversely, we cannot release faith that we have not yet developed.**

To increase our personal spiritual momentum requires that we first take steps to build our faith. Faith does not usually come accidentally into the life of a believer. It must be intentionally developed.

Several steps are spelled out in Scripture to help us build our faith. Again, the predominate means to build our faith is spelled out in the Book of Romans:

Rom 10:17 - "So then faith cometh by hearing, and hearing by the Word of God."

Many books have been written on how to build our faith so this book is not intended to be an in-depth instruction manual for this. However, we will bring out a few points about developing faith to aid in our instruction on spiritual momentum.

Time spent reading, speaking, and meditating God's Word all work to build our faith in His promises. The more that we permit Scriptural promises to roll over and over again in our thinking allows the Word to transition from residing in our head to being solidified within our hearts. Once the Word takes position within our hearts, a foundation for faith in our targeted promises of God has been established.

Another endeavor that will help to bolster our faith is *practice*. As we step out and use the faith that we have, God will show Himself faithful to His Word. Such experience allows us to launch out in even higher use of our faith thus allowing us to go from *faith to faith*.

As well, *church attendance*, listening to *faith-based Bible teaching* via television, the internet, or other media all may work to build our faith. The Christian hungry to build his faith will pursue exposure to the Word through whatever means that he can.

There remains another key step that believers may take to enhance their faith. In fact, by combining this activity with time in the Word, faith development will be accelerated. This faith-boosting step is to spend extensive time praying in the Spirit. For those who have been baptized in the Holy Spirit, God has equipped them with a phenomenal tool to accelerate their faith growth. Those baptized in the spirit have been given their own personal prayer language that allows them to pray in tongues.

Jude informs us that this act helps to build our faith.

*Jude 1:20 - "But ye, beloved, **building up yourselves on your most holy faith**, praying in the Holy Ghost,"*

Scripture speaks to us about praying in tongues in 1Cor-inthians:

*1Cor 14:2 - "For he that speaketh in an unknown tongue speaketh not unto men, but unto God: for no man understandeth him; howbeit in the spirit **he speaketh mysteries**."*

However, Scripture also tells us that God always rewards us for what we sow.

*Gal 6:7 - "Be not deceived; God is not mocked: for **whatsoever a man soweth**, that shall he also reap."*

As we pray in tongues, we are releasing mysteries back unto God. But because God is not mocked, as we do so and then begin to read the Word of God, it will come back to us in the form of revelation knowledge. God will reveal greater and greater depths of His Word to us as we spend time praying in tongues. Then, as we meditate upon this revelation, faith arises to all-new levels.

Again, we cannot expend faith that we have not developed and fueled anymore than we can drive a car without gasoline. Therefore, all of these faith-building activities work to fill our spiritual fuel tanks with the raw fuel of faith. Only then can we release that faith to develop the spiritual momentum that we desperately need to live above the curse of sickness and other oppressive forces.

It is interesting to note from our formula for momentum the impact that increasing our faith and faith application time has upon our spiritual momentum. If we double either our faith level or our application time, our spiritual momentum will also double. However, since the product of force and time equals momentum, if we double both faith and our application time, our spiritual momentum will increase fourfold. What a great opportunity we have to trample the enemy under our feet through the application of these principles.

Hopefully, by now you are seeing the importance of developing spiritual momentum in your life.

CHAPTER 4

THE WORKS OF GOD

In physics, work is defined as *a force applied over a distance*. The formula for this work is:

Work (W) = Fd

where F equals the force applied and d equals the distance the object is moved.

If you push against a parked car for an hour but it does not move, since the distance the car is moved equals zero, no work is actually done. However, imagine instead that the car is on a level surface and is placed in neutral. When we now begin to push against it, as the car begins to move, work is being accomplished. The total work completed is equal to the product of the force that we exerted to push the car times the distance that it is moved.

For example, if we push against the car with a continued force of 100 pounds such that it travels 100 feet, then we have completed 10,000 foot-pounds of work.

Work to Push a Car 100 Ft.

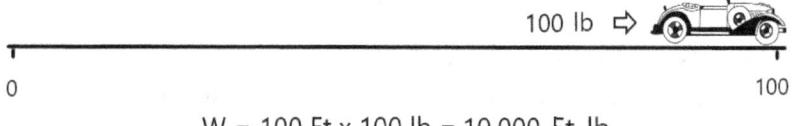

W = 100 Ft x 100 lb = 10,000 Ft-lb

Notice in this formula for **work** that **mass (m)** does not appear. The formula for work does not consider the size of the object being moved. It only focuses on how far the object travels. For example, if we apply the same 100 pounds of force to move a bicycle 100 feet, the same 10,000 foot-pounds of work is accomplished. The difference of course is that the bicycle travels the 100 feet in a much shorter time.

Work to Push a Bicycle 100 Ft.

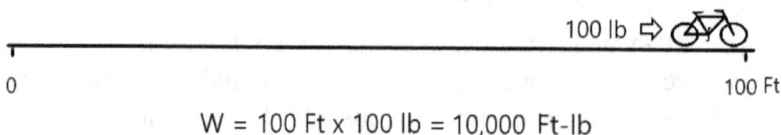

W = 100 Ft x 100 lb = 10,000 Ft-lb

So, work is dependent upon how much force that we apply and how far the object is moved as a result of that force. If we slowly lift a 10-pound barbell over our head 10 times, we do the same amount of work as slowly lifting a 100-pound barbell over our head 1 time. (Slowly lifting the barbells assures that the force applied closely approximates the weight of the barbells.)

However, imagine that more weight is placed on the barbells than we can lift off of the floor. Regardless of the amount of force that we exert, if the barbells do not move, no work is accomplished.

Spiritual Work

But how might this concept of work in physics apply to us spiritually? First understand, from Scripture, we know that God places much emphasis upon the works that we do for Him. In fact, the word *work* appears in the King James Bible in one form or another 689 times.

Yet we also know from the Scriptures that no amount of good

works is adequate to earn salvation:

*Gal 2:16 - "Knowing that a man is **not justified by the works of the law**, but by the faith of Jesus Christ, even we have believed in Jesus Christ, that we might be justified by the faith of Christ, and **not by the works of the law**: for by the works of the law shall no flesh be justified."*

However, once we are saved, God expects us to target producing good works for His Kingdom and glory.

*Mat 5:16 - "Let your light shine before men, that they may see your good **works**, and glorify your Father which is in heaven."*

Even Jesus declared that He came to perform the works assigned to Him of the Father.

*John 4:34 - "Jesus saith unto them, My meat is to do the will of him that sent me, and to finish his **work**."*

Jesus even declared that the Father works.

*John 5:17 - "But Jesus answered them, My Father **worketh** hitherto, and I work.*

God has ordained the five-fold ministry to train the Church to do the works of God.

*Eph 4:11-12 - "And he gave some, apostles; and some, prophets; and some, evangelists; and some, pastors and teachers; For the perfecting of the saints, for the **work** of the ministry, for the edifying of the body of Christ:*

We also discover in Scripture that God rewards us based on our works.

*Psa 62:12 - "God rewards every man according to his **work**."*

*Ecc 12:14 - "For God shall bring every **work** into judgment, with every secret thing, whether it be good, or whether it be evil."*

Mat 16:27 - "For the Son of man shall come in the glory of his

Father with his angels; and then he shall reward every man according to his **works**."

Heb 6:10 - "For God is not unrighteous to forget your **work** and labour of love, which ye have shewed toward his name, in that ye have ministered to the saints, and do minister."

1Pet 1:17 - "And if ye call on the Father, who without respect of persons judgeth according to every man's **work**, pass the time of your sojourning here in fear:"

I understand that numerous Bible verses are listed here. However, the purpose is to show just how much emphasis God places upon the works that we accomplish for Him once we are saved. But what can we determine from our physics formula for work? Recall that:

$$W = Fd$$

Also remember that for our spiritual computations, **F** equates to the Heavenly forces accessed and released by *faith*. And, since we are using our faith to move obstacles from our path, then **d** represents the *distance mountains are moved* from before us. We can use our faith to move the mountain a few inches or we can have it thrown into the sea. The work that we accomplish is equal to the magnitude of our faith applied multiplied times the level of change that we experience.

For example, if we are praying for the healing of our bodies, have we maintained our faith pressure until all of the symptoms are totally obliterated? Or, have we ceased from applying our faith once we were only slightly better? If we are praying for finances, did we maintain our faith pressure against lack merely to see a few bills paid? Or, did we develop and apply our faith until all debt was removed from our accounts and we experienced financial overflow?

The key factor in completing these spiritual works for God is that, per our formula, we can do no works for God without faith. We see this tie between completing any works for God and using our faith in the Bible.

*John 6:28-29 - "Then said they unto him, What shall we do, that we might **work** the **works** of God. Jesus answered and said unto them, This is the **work** of God, that ye **believe** (have faith) on him whom he hath sent."*

*1Thes 3:10 - "Remembering without ceasing your **work of faith**, and labour of love, and patience of hope in our Lord Jesus Christ, in the sight of God and our Father;"*

*Mat 13:58 – "And he did not many mighty **works** there* (in His hometown) *because of their **unbelief** (lack of faith)."*

In fact, James, the brother of Jesus, lets us know that if we are doing no works for God then our faith is either non-productive or non-existent.

*Jas 2:17 - "Even so **faith, if it hath not works**, is dead, being alone."*

*Jas 2:26 - "For as the body without the spirit is dead, so **faith without works is dead** also."*

If we have zero faith to function in the supernatural power of God, then we will not be able to complete any spiritual work for Him. This also holds true even if we may have developed faith in God's supernatural power but have not yet applied any of it. In such cases, our faith in salvation may get us to heaven but, once there, we may discover that we have no rewards accredited to our accounts.

*1Cor 3:13-15 - "Every man's **work** shall be made manifest: for the day shall declare it, because it shall be revealed by fire; and the fire shall try every man's **work** of what sort it is. If any man's **work** abide which he hath built thereupon, he shall receive a reward. If any man's **work** shall be burned, he shall suffer loss: but he himself shall be saved; yet so as by fire."*

**A FIFTH LAW OF KINGDOM TRANSFORMATION –
Without faith, we can do no mighty works for God.**

Another interesting point to keep in mind is that, in our formula for *work*, the size of the mountain does not noticeably enter into the equation. It is not always necessary that we use our faith to move enormous mountains. What is important is that we use the faith that we do have to begin moving something. So many Christians who learn about faith often declare right away how they want to conquer every problem in the world. However, we must learn to believe God to heal a headache or pay the light bill before we can use our faith to raise the dead or build a major church structure. At least then we will have moved something out of the way and will have accomplished some level of work for God.

Another point to be made from our study of spiritual work is that we must be realistic in what we target to accomplish for God by our faith. It is always admirable for the Christian to target doing great works for God. Through such mindsets, we are challenged to stretch our faith to higher and higher levels. However, it is not wise to allow spiritual pride to affect us such that we attempt to move Mt. Everest on mole-hill faith. If the mountain that we are attempting to move is too large for us then no genuine spiritual work will be completed.

A SIXTH LAW OF KINGDOM TRANSFORMATION –
Until something is actually altered or transformed supernaturally by faith in God's Word, no Kingdom work has actually been accomplished.

As end-time laborers for God, we will be held accountable for the works that we do with the gifts and talents that we have been given. In Scripture, in both the *Parable of the Talents* and the *Parable of the Pounds*, each man was rewarded according to his works. However, also in both cases, those who did nothing with their God-given initial investments were cast into outer darkness.

In perhaps the most amazing promise of the Bible, Jesus relates to us the tie between the works that we can accomplish for God and our faith in the Word.

*John 14:12 - "Verily, verily, I say unto you, He that **believeth** on me, the **works** that I do shall he do also; and greater **works** than* *these shall he do; because I go unto my Father."*

Here Jesus declares that if we will believe on Him, and subsequently His Word, then nothing shall be impossible to us. Even now, those who comprehend and have developed faith to operate in the supernatural power of God are witnessing miracles in their lives as in no prior generation.

To maximize our spiritual work output, we must be determined to utilize our faith for every aspect of our lives. When we, by faith, displace sickness, lack, strife, and anger far from us, we accomplish work for God in multiple areas simultaneously.

CHAPTER 5

DIVINE ENERGY

The Anointing

In this chapter, we are going to examine some laws of physics that tie the motion of an object with the energy that it possesses. From this, we will discover key steps that we can take to enhance our ability to walk in the energy that proceeds forth from Heaven itself. There is a divine energy that God has made available to us – *the anointing*. The anointing is the energy that emanates from the Kingdom of Heaven which allows us to function in the supernatural power of God. It is this anointing to which Jesus attributed His ability to operate in miracle power.

> *Luke 4:18-19 - "The Spirit of the Lord is upon me, because he hath anointed me to preach the gospel to the poor; he hath sent me to heal the brokenhearted, to preach deliverance to the captives, and recovering of sight to the blind, to set at liberty them that are bruised, To preach the acceptable year of the Lord."*

The anointing is the source of power that enabled Jesus, then later His disciples, and now us to perform the impossible. In Isaiah, the anointing is declared able to free us from any of the restraints of the enemy.

> *Isa 10:27 - "And it shall come to pass in that day, that his burden shall be taken away from off thy shoulder, and his yoke from off thy neck, and the yoke shall be destroyed because of the anointing."*

This divine spiritual energy that Scripture refers to as the anointing is the source of power to heal the sick, cast out demons, and raise the dead. It is the heavenly energy that enables us to possess supernatural strength, joy, and even love. Every Christian should set their sights on enhancing the anointing in which they walk. But what can the formulas of physics teach us about this energy that proceeds from the throne of God?

Potential Energy

We have limited our study of physics to the area of linear motion – motion in a straight line. There are two predominant forms of energy that pertain to linear motion, *potential energy,* and *kinetic energy*.

In our last chapter, we discussed the topic of *work* from both a natural and a spiritual standpoint. A principle of physics that is now key for us to understand is that **energy provides the ability to do work**.

From a physics standpoint, **work and energy are interchangeable** – either one can be used to produce the other. In fact, work can be used to create both potential energy and kinetic energy. Conversely, both potential energy and kinetic energy can be used to accomplish work.

In physics, the potential energy of an object represents the amount of work that it can produce by moving from a higher elevation to a lower one. Hydroelectric power produced by many dams is the result of capturing the energy difference between the higher potential energy of water at the top of the dam versus the lower potential energy at the foot.

A simplified formula for potential energy, which we will designate as **PE**, is:

$$PE = W_T h$$

where W_T equals the weight of the item and h equals its height or elevation.

Weight is the result of the force of gravity acting upon a mass. It is this force of weight that causes the bathroom scales to often read higher values than many of us would prefer. What I want you to grasp from this formula is that **weight is a force!** As we elevate an object to greater and greater elevations, the object increases in potential energy. This potential energy qualifies the object to produce work at some point in the future.

It is interesting to note that our formula for work is **W = Fd** - a force times a distance. Because *weight is a force* and *height is a distance* and **PE = W$_T$ h**, we can see the direct tie between potential energy and its ability to do work.

Spiritual Potential Energy

But how does this analysis of potential energy relate to our study of spiritual physics? As we have mentioned at the outset of this chapter, God has made available to us an energy source Scripture refers to as the *anointing*.

As just mentioned, *weight is a force*. In our examination of *spiritual physics*, **we have discovered that faith is the key to activating supernatural *forces* from *God's Kingdom*.** As well, we have learned that potential energy is the result of elevating the weight of an object to higher and higher levels.

Since both weight and faith represent the ability to apply force, we can ascertain that our *spiritual potential energy* is increased as we raise and fuel our faith to higher and higher levels. As we take time to elevate our faith and anointing reserves, we increase our capacity to accomplish supernatural feats for God. Time in the Word of God, time in prayer, extended time praying in tongues, etc. all work to increase and fuel our faith and thus our spiritual potential energy.

Again, you cannot expend faith that you do not have. As well, you cannot accomplish any genuine work for God's Kingdom unless you have first built up your spiritual potential energy reserves. The disciplined effort to build ourselves up in God fills our spiritual fuel tanks with the raw energy of the anointing of the Holy Spirit. Should we neglect these preparatory activities, our tanks become low, flesh arises, and we begin to attempt to serve God out of our own efforts and strength.

A SEVENTH LAW OF KINGDOM TRANSFORMATION – The believer may build up his internal anointing reserves which allow him to later perform supernatural works for God.

The Bible lets us know that we can continually build our faith to ever-increasing levels.

> *Rom 1:16-17 - "For I am not ashamed of the gospel of Christ: for it is the power of God unto salvation to every one that believeth; to the Jew first, and also to the Greek. For therein is the righteousness of God revealed **from faith to faith**: as it is written, The just shall live by faith."*

As we build and fuel our faith to higher levels, our stored potential anointing increases as well. Every responsible Christian should be willing to re-prioritize their days to take time to build up their spiritual potential energy reserves. Only then are we equipped to perform the miraculous on demand.

Too many ministers focus singularly on preparing Bible messages and not enough time filling their spiritual fuel tanks. Thus, they lack any anointing to function in the supernatural power of God.

But of course, energy in the form of potential energy is not yet producing any work. To accomplish any results requires that our potential energy be activated or put in motion.

Kinetic Energy

Our Second type of energy that we want to examine is *kinetic energy*. Kinetic energy is the energy that an object possesses due to its being in motion. Recalling that work and energy are interchangeable, to derive our formula for kinetic energy, we need to begin with our formula for work.

$$W = Fd$$

To convert our work formula to an equation that reflects motion, we need to find the impact that *velocity* has on the components of our work formula, **Force,** and **distance**. The formula that determines how far an object travels is:

$$d = v_{avg}\, t$$

where v_{avg} equals average velocity and t equals the time the object has been in motion.

Assuming a constant acceleration, the formula for average velocity is:

$$v_{avg} = v/2$$

where v equals the objects *final velocity*.

Therefore, we can revise our distance formula to:

$$d = vt/2$$

By substituting this into our formula for work we get:

$$W = Fvt/2$$

We are now beginning to convert our formula for work to an equation reflecting motion. However, the force that we apply also has an impact on the velocity of an object. Recalling our formula for force:

$$F = ma$$

where a equals the acceleration of the object.

Also recalling that $\quad a = v/t$

By substituting into our equation for force we obtain:

$$F = mv/t$$

If we now substitute this formula for force into our revised formula for work we obtain:

$$W = mv/t \quad \text{times} \quad vt/2$$

or
$$W = mv^2t/2t$$

I'm sure that some may not comprehend all of the mathematics and physics covered in these steps. However, algebraically the **t**'s cancel each other out and we thus obtain a simplified equation for work:

$$W = mv^2/2 \quad \text{or} \quad W = \tfrac{1}{2} mv^2$$

We have now converted our formula for work to an equation that reflects the motion of an object. This subsequently becomes our equation for **kinetic energy** (**KE**).

$$KE = \tfrac{1}{2} mv^2$$

From this, we can see that kinetic energy is directly proportional to the mass of an object and increases exponentially with its velocity. In other words, as with momentum, the more an object weighs and the faster that it travels - the more kinetic energy it possesses. Also recall that the velocity of an object is the result of a force that has been applied to that object. So then, force plays a role in building the kinetic energy of a mass.

We have now shown that energy can be converted into work and work can be used to produce energy. This becomes a key to every aspect of our society. As mentioned before, hydroelectric dams convert the potential energy of water at one elevation to kinetic energy as it falls to a lower elevation. This kinetic energy of the water is used to produce work as it spins turbines at the bottom of the dam. This work done by the turbines is then converted to electrical energy that powers our homes.

As well, the potential energy inherent in gasoline is converted to mechanical work by our automobile engines which enables our car to travel down the road; thus possessing kinetic energy.

There are multiplied other examples both from industry and nature that demonstrate the interchangeability of work and energy. This becomes an interesting facet in our analysis as we continue our study of spiritual physics.

Potential Energy Being Converted to Kinetic Energy then to Electrical Energy

Spiritual Kinetic Energy

We have discussed how work and energy are interchangeable. But how might this apply to us spiritually? Recall that our spiritual energy reflects the supernatural anointing of God both upon and flowing through us. Therefore, apparently the works that we can do for God are dependent upon the anointing that we develop and learn to impart.

The significance of this is monumental. Our *anointing* - accessed by faith, allows us to do the *works of God*. Without this supernatural empowerment of the anointing, all of our efforts to perform works for God are produced out of our own flesh. Such efforts to do good works for God birthed from flesh rather than the anointing become wood, hay, and stubble and will be burned.

Where spiritual potential energy represents built up and stored personal anointing in the life of the believer, spiritual kinetic energy is the result of that anointing being activated or placed in motion. This *released spiritual energy* produces a *flow of anointing* that allows us to operate in supernatural power. It is this *anointing* that allows us to do the works of God.

While faith that has been built up has the potential to move mountains, until it is released through speaking faith-filled words, it remains untapped. However, spiritual potential energy, once released, is converted to spiritual kinetic energy. It is this *spiritual energy in motion* that allows us to blast through symptoms of sickness, lack, and oppression.

Scripture also lets us know that potential spiritual energy can be released into the lives of others through physical contact. In the Gospels, we repeatedly see Jesus heal the sick through laying His hands on them.

Luke 4:40 - "Now when the sun was setting, all they that had any sick with divers diseases brought them unto him; and **he laid his hands on every one of them**, *and healed them."*

Jesus has also instructed us to lay hands on the sick.

Mark 16:17-18 - "And these signs shall follow them that believe; In my name shall they cast out devils; they shall speak with new tongues; They shall take up serpents; and if they drink any deadly thing, it shall not hurt them; **they shall lay hands on the sick, and they shall recover.**"

In fact, the writer of Hebrews even referred to the laying on of hands as a basic foundation block of Christianity.

Heb 6:1-2 - "Therefore leaving the principles of the doctrine of Christ, let us go on unto perfection; not laying again the foundation of repentance from dead works, and of faith toward God, of the doctrine of baptisms, and of **laying on of hands**, *and of resurrection of the dead, and of eternal judgment,*

It is no wonder that the act of laying hands on the sick is so opposed by the enemy. It should be obvious that this action is key to transferring the anointing to those in need of healing. The act of laying hands on someone is for more than only releasing healing anointing into someone's life. The act of laying hands on other believers may be used to impart spiritual gifts and even ordain leaders into office.

As we go forth to serve God by faith, the internal potential energy of our personal anointing may be converted to kinetic spiritual energy in the form of released anointing. This released anointing accomplishes supernatural works for the Kingdom. These supernatural works transform the lives of those to whom we minister.

Therefore, it becomes vital for the individual who wants to continually accomplish the *greater works* that Jesus spoke of to keep himself filled with the energy of God's anointing. As with faith, this anointing can only be maintained through spending extensive time in God's Word and the worship of Him.

It is not enough just to study and believe the Word of God. To actually produce any work for God requires that we activate (or put in motion) our faith. Until then, our anointing remains dormant. We see this need to convert potential faith to operational faith in the book of James.

Jas 2:17 - "Even so faith, if it hath not works is dead being alone."

Obviously, until we do something productive with our faith, no work for God is accomplished.

Through a closer look at our formula for kinetic energy, we can see another interesting aspect that has spiritual implications for the Christian. Our formula for kinetic energy is:

$$KE = \tfrac{1}{2} mv^2$$

this may be rewritten as

$$KE = \tfrac{1}{2} (mv)v$$

Recall that **mv** equals *momentum*. Therefore, from a physics standpoint, if an object has kinetic energy then it also possesses momentum. But how does this apply to us spiritually?

Remember that we have already found that a Christian with great spiritual momentum can bowl over any attack of the enemy. This revised formula lets us know that this ability to overcome any obstacle through spiritual momentum is actually supplied by the anointing. It is not via any strength or ability of ourselves that we may trample over the enemy's best efforts to hinder us. It is through the supernatural energy of the Kingdom of God – the *anointing*.

In sports and even in the business world, there is much talk today about the value of carrying 'momentum'. Sports teams know that if they can somehow grasp and carry momentum during a competition, they are much more likely to win the game. However, this type of momentum that the world strives for is not the same as that which God's Kingdom offers.

The momentum of the world is based on excitement and confidence. This earthly momentum can accomplish much in the world but it contains no capacity to produce supernatural results. We, as believers, need to access the power from heaven that enables us to accomplish the impossible. The anointing inherent in the life of the believer provides spiritual momentum that has the capacity to perform what the world cannot.

How sad it is that much of the Church today cannot distinguish between worldly momentum and that supernaturally produced by the anointing. Because of this, energetic entertainment venues have replaced truly anointed church activities. However, with God, humanly produced enthusiasm is not a suitable substitute for that only Heaven will release via faith.

To summarize what we have just discussed, if we have no *faith*, then we will be void of *anointing*. And, if we have no *anointing*, we can do no mighty *works* for God. It is these *works* by *faith* that allow us to please God.

> *Heb 11:6 - "For without faith it is impossible to please him: for he that cometh to God must believe that he is, and that he is a rewarder of them that diligently seek him."*

CHAPTER 6

THE POWER OF GOD

In physics, *power* is defined as *the amount of work produced per unit of time*. The formula for this is:

$$P = W/t$$

where P equals Power, W equals Work, and t equals time.

Knowing that **Work** equals **Force** x **distance**, then Power may be rewritten as:

$$P = Fd/t$$

Common terms often used to quantify Power are the **Horsepower** and the **Watt**. We most often think of horsepower when referring to mechanical power and the watt when referring to electrical power. Originally, the horsepower was meant to approximate the amount of work that a draft horse could perform. The horsepower became defined as 550 foot-pounds/second. This means that one horsepower can lift a 550-pound weight a distance of one foot for each second of operation.

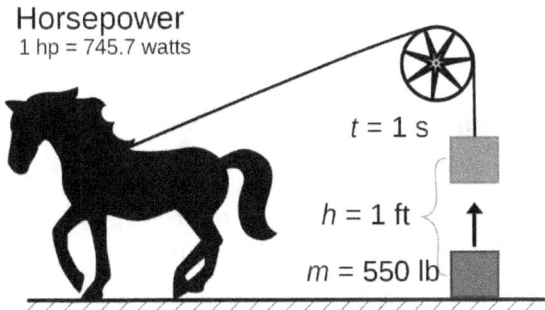

In the imperial system of measurement, one horsepower equals 745.7 watts.

When we speak of the horsepower of an engine, we are describing the work producing capability of that engine. Large horsepower engines can produce more work in a set period of time than small horsepower engines.

We have also discussed how work can be converted to energy. Therefore, an alternative formula for power may be written as:

$$P = (½ mv^2)/t$$

where ½ mv² equals kinetic energy

This formula may be simplified to:

$$P = mv^2/2t$$

Most engines that we are familiar with convert potential energy in the form of fuel to work or kinetic energy as with a vehicle accelerating down the road. The power (horsepower) rating of that engine indicates how fast the potential energy can be converted to useful work or energy in some other form. Therefore, if we want to accomplish high levels of mechanical work in shortened amounts of time, it is vital that our equipment be *powerful*. Large bulldozers can move much more earth than small ones. As well, race car drivers look to maximize the power output of their car's engines if they want to effectively compete.

But how does the physics topic of power apply to us spiritually? Jesus said:

*Acts 1:8 - "But ye shall receive **power**, after that the Holy Ghost is come upon you: ... ."*

From Scripture, we see the vital importance of walking in great spiritual power! First know that there are two words in the New Testament most commonly translated as power. These are ***exousia*** and ***dunamis***.

The Greek word *exousia* represents *authority*. In our case, *exousia* infers the *spiritual authority* that we have access to as believers. The word *exousia* is used in our New Testament 103 times. It is this authority that Jesus gave to his disciples to cast out demonic spirits.

> *Mat 10:1 – "And when he* (Jesus) *had called unto him his twelve disciples, he gave them power* **(exousia)** *against unclean spirits, to cast them out, and to heal all manner of sickness and all manner of disease."*

> *Luk 10:19 - "Behold, I give you power* **(exousia)** *to tread on serpents and scorpions, and over all the power* (dunamis) *of the enemy: and nothing shall by any means hurt you."*

However, the Greek word for power that we want to focus on is the word *dunamis*. *Dunamis* refers to the power that an individual or thing may put forth. *Dunamis* is often described from a Biblical standpoint as the spiritual force that is able to back up our *exousia* authority. From Scripture, we see that Jesus walked in both *exousia* and *dunamis*.

> *Luk 4:36 - "And they were all amazed, and spake among themselves, saying, What a word is this! for with authority* **(exousia)** *and power* **(dunamis)** *he* (Jesus) *commandeth the unclean spirits, and they come out."*

In fact, the miracles that Jesus performed were accom-plished through these two sources of power.

In some instances, *dunamis* even indicates the power or ability to perform miracles. We see in First Corinthians that the G*ift of the Spirit* listed to perform miracles is actually the ability to walk in *dunamis*.

> *1Cor 12:10 - "To another the working of miracles* **(dunamis)**;..."

It should be obvious to us that we need to walk in this *dunamis* spiritual power. It is interesting from Luke 10:19 (above), that our *exousia* alone is greater than the enemy's *dunamis*. How much more, the *dunamis* of the believer is able to destroy all of the works of the devil.

We also see from Scripture that believers should be operating in this same *dunamis* power. Paul declared:

1Cor 2:4 - "And my speech and my preaching was not with enticing words of man's wisdom, but in demonstration of the Spirit and of power **(dunamis)**.*"*

Rom 15:19 - "Through mighty **(dunamis)** *signs and wonders, by the power* **(dunamis)** *of the Spirit of God; so that from Jerusalem, and round about unto Illyricum, I have **fully preached** the gospel of Christ."*

Apparently, until miracle *dunamis* power is on display in a region, the Gospel has not yet been fully preached there.

The word *dunamis* is even directly tied to the *flow* of spiritual power from Jesus. In the story from Mark where the woman was healed of an issue of blood, spiritual power was released when she touched the hem of His garment. Jesus was so familiar with the power that was upon His life that He recognized the moment that this took place.

Mark 5:30 - "And Jesus, immediately knowing that **virtue (dunamis)** **had gone out of him**, *turned him about in the press, and said, Who touched my clothes?"*

It is therefore obvious from Scripture that the dunamis spiritual power that we walk in is directly correlated to the anointing that we carry. Therefore, d*unamis* indicates the ability of the believer to flow in spiritual energy. As we discussed in our chapter on Energy, this spiritual energy is known as the *anointing*. **In fact, we can think of our *spiritual power* as the *rate of anointing* that can flow through us.** We see that the Bible again ties the dunamis power in which Jesus walked to the anointing.

*Acts 10:38 - " How God **anointed** Jesus of Nazareth with the Holy Ghost and with **power** (dunamis): who went about doing good, and healing all that were oppressed of the devil; for*

God was with him."

Christians who develop themselves to walk in high levels of spiritual power become equipped to function in the miraculous. Therefore, for the Christian to operate in any high level of anointing, it is imperative that they develop and increase their personal level of spiritual power.

But what do our formulas for power relate to us? From our prior formula, we discovered that:

$$P = Fd/t$$

where P equals our spiritual power

Let's take a look at how each of the variables in this equation impacts our spiritual power level.

The Relationship between FAITH and Spiritual Power

Recall from our prior analysis, that our force **(F)** equates to the forces of Heaven activated by faith. Therefore, as with every other spiritual attribute that we have investigated, spiritual power is dependent upon the faith that we will develop in our ability to carry this power. We see the tie between faith and dunamis power displayed in numerous Scriptures.

*Acts 6:8 - "And Stephen, full of **faith and power** (dunamis), did great wonders and miracles among the people."*

*1Cor 2:5 - "That your **faith** should not stand in the wisdom of men, but in the **power** (dunamis) of God.*

*1 Pet 1:5 - "Who are kept by the **power** (dunamis) of God through **faith** unto salvation ready to be revealed in the last time."*

A key to walking in great spiritual power is to target using our faith to access *dunamis*. The hungry Christian should let God know that they desire to walk in this power. We may accelerate this process as we fast and pray to prepare ourselves to carry more of it. The believer who desires to carry the full power of God will fervently seek to walk in the *Gifts of the Spirit* as listed in First Corinthians Chapter 12.

The Relationship between TIME and Spiritual Power

One of the most important things that a Christian should desire is to carry high levels of spiritual power. We mentioned how a large bulldozer can move much more earth than a small one in the same amount of time. In the same fashion, to move large mountains in a reasonable period of time requires high levels of spiritual power. If the believer does not make an effort to increase in spiritual power, then many of the obstacles that would oppose him may remain unmoved.

However, as we develop our capacity to carry this spiritual power, we gain the ability to operate in the miraculous.

Our formula for power relates to us the amount of work that can be accomplished for a unit of time. Therefore, if the amount of work that needs to be completed remains fixed (as with moving a mountain or other obstacle), if the believer possesses great power, he may accomplish the task in a far shorter time than if he possesses little spiritual power. From this we can see that the individual with great spiritual power will experience accelerated manifestation in his life.

AN EIGHTH LAW OF KINGDOM TRANSFORMATION – The Christian's ability to *quickly* exercise dominion over all of the powers of darkness is directly tied to his level of spiritual power.

The Relationship between DISTANCE and Spiritual Power

In our above formula for power, **d** is the distance that an object moves due to the force that is applied to it. However, the term distance may seem to have little meaning to us from a spiritual standpoint. When we are contemplating things that are taking place in an unseen realm, what could distance imply? To answer this, think again how Jesus stated in Mark that our faith could remove mountains.

> *Mark 11:23 - "For verily I say unto you, That whosoever shall say unto this mountain, Be thou removed, and be thou cast into the sea; and shall not doubt in his heart, but shall believe that those things which he saith shall come to pass, he shall have whatsoever he saith."*

From this, we can define **d** in our formula as the distance that the mountain is moved from before us. The implications of this are substantial – as we operate in higher and higher levels of spiritual power, adverse circumstances may be repositioned far from us. We see this demonstrated in a Scripture that promises us not just divine healing but also divine health.

> *Psa 91:10 - "There shall no evil befall thee, **neither shall any plague come nigh thy dwelling.**"*

This psalm declares that sickness and disease can be placed far from us! What a marvelous promise of God. Evidently, it is possible for saints to never need to be sick another day of their life. However, this promise is given to one who *dwells* in the secret place with God. Obviously, this person is taking dramatic steps to enhance his level of spiritual power. We see another amazing promise in this same psalm.

> *Psa 91:7 - "A thousand shall fall at thy side, and ten thousand at thy right hand; **but it shall not come nigh thee.**"*

Such a phenomenal promise certainly belongs to one who has increased their ability to carry the power of God.

**A NINTH LAW OF SPIRITUAL TRANSFORMATION –
Maximized spiritual power enables the Christian to live far above every attack of the enemy such that the devil's assaults cannot even touch him.**

Spiritual Power is Key

In this book, we have covered the topics of Acceleration, Momentum, Work, Energy, and now Power. From what we have discovered from a spiritual standpoint, all of the above attributes require the application of faith in the promises of God. However, what we also find from a closer look at our spiritual physics formulas is that all of the previous qualities we covered can all be accessed through enhanced spiritual power. Looking again at our formula for power we find:

$$P = W/t \quad \text{where W equals } \textit{Work}$$

From this we see that if we can develop the ability to walk in great spiritual power, then we are equipped to accomplish much work for God. Looking at the formula from another viewpoint:

$$P = Fd/t \quad \text{where F equals } \textit{Faith}$$

Therefore, if we walk in great power, we will be able to release faith to move large mountains. Also:

$$P = \tfrac{1}{2} mv^2/t \quad \text{where } \tfrac{1}{2} mv^2 \text{ equals } \textit{Energy}$$

If we will develop spiritual power, then we will be anointed with spiritual energy which enables us to accomplish the miraculous! And finally:

$$P = mv^2/2t \quad \text{or rewritten,} \quad P = (mv)v/2t$$
In this case, mv equals Momentum

So then, if we will develop spiritual power, we will have the ability to create great spiritual momentum in our lives.

What each of these demonstrates is that if we will take active steps to increase our spiritual power capacity, then all of the other spiritual principles discussed in this book become easier to manifest.

A TENTH LAW OF SPIRITUAL TRANSFORMATION –
Increased capacity to carry spiritual power enhances:

1) **Our ability to accomplish much WORK for God,**
2) **Our FAITH to move mountains,**
3) **Our spiritual ENERGY or anointing level, and**
4) **Our ability to produce elevated spiritual MOMENTUM allowing us to bowl over every attack of the enemy.**

Increasing our Spiritual Power Capacity

So then the question arises, "How do I increase my ability to walk in the power of God?" Of course the answer is tied to our desire to seek Him, know Him, abide with Him, and our desire to demonstrate His power. Only through a deep relationship with God should we be entrusted with such capability.

As we have already alluded, prayer and fasting accelerate our spiritual development to carry the power of God. Much time in God's Word is also mandatory.

However, there is another key to walk in enhanced supernatural ability. Our initial verse quoted in this chapter holds a key. As Jesus was about to ascend to the Father, He instructed the disciples to wait in Jerusalem until they were ***Baptized in the Holy Ghost***. Regarding this baptism, Jesus declared:

Acts 1:8 - *"But ye shall receive **power**, after that the Holy Ghost is come upon you: ... ."*

Jesus instructed the disciples to wait for this baptism because it was going to become a major key for them to walk in supernatural power. For those who want to walk in the miraculous, it becomes paramount to seek this free gift available to all who will ask.

Once the believer is baptized with the Holy Spirit, they are equipped with a supernatural prayer language that allows them to always pray the perfect will of God. Such prayers allow God to correct, develop, anoint, and empower us in an accelerated fashion.

Through using this prayer language, Christians are equipped to develop at the maximum rate possible.

Life's priorities all shift for those who really desire to carry the power of God to open blind eyes, cast our demons, and raise the dead.

However, someone might say: "I really am not interested in walking in miracle power. I am only focused on walking in love!" Please understand that the genuine *agape* love of God is supernatural in content. In fact, the true love of God in action takes the form of an anointing. Since the anointing is a form of spiritual energy, if we genuinely want to walk in high levels of love, then we will also need to seek to walk in high levels of spiritual power.

For more on the topic of supernatural love, please read my book entitled "THE SUPERNATURAL POWER OF LOVE."

CHAPTER 7

CONCLUSIVE RESULTS

This exercise in comparing the natural laws of physics to the spiritual laws of the Kingdom of God is, of course, only theoretical. However, hopefully through the formulas shown, you have been able to recognize some of the basic components necessary to live by faith and walk in supernatural power.

To review, listed below are ten **Laws of Spiritual Transformation** that we derived from our studies in this book.

**THE FIRST LAW OF KINGDOM TRANSFORMATION -
No situation or circumstance on earth will be altered by the supernatural power of God until someone releases *force* through the operation of *faith* in God's Word.**

**A SECOND LAW OF KINGDOM TRANSFORMATION –
As our faith increases, we can move larger and larger mountains in shorter amounts of time.**

**A THIRD LAW OF KINGDOM TRANSFORMATION -
Spiritual momentum enables the believer to push back and overcome every attack of the enemy.**

**A FOURTH LAW OF KINGDOM TRANSFORMATION –
As we spend extended time building our faith and confessing God's Word, we build spiritual momentum.**

**A FIFTH LAW OF KINGDOM TRANSFORMATION –
Without faith, we can do no mighty works for God.**

**A SIXTH LAW OF KINGDOM TRANSFORMATION –
Until something is actually altered or transformed supernaturally by faith in God's Word, no Kingdom work has actually been accomplished.**

A SEVENTH LAW OF KINGDOM TRANSFORMATION –
The believer may build up his internal anointing reserves which allow him to later perform supernatural works for God.

AN EIGHTH LAW OF KINGDOM TRANSFORMATION –
The Christian's ability to *quickly* exercise dominion over all of the powers of darkness is directly tied to his level of spiritual power.

A NINTH LAW OF SPIRITUAL TRANSFORMATION –
Maximized spiritual power enables the Christian to live far above every attack of the enemy such that the devil's assaults cannot even touch him.

A TENTH LAW OF SPIRITUAL TRANSFORMATION –
Increased capacity to carry spiritual power enhances:

1) **Our ability to accomplish much WORK for God,**
2) **Our FAITH to move mountains,**
3) **Our spiritual ENERGY or anointing level, and**
4) **Our ability to produce elevated spiritual MOMENTUM allowing us to bowl over every attack of the enemy.**

Please do not attempt to convert this list of spiritual principles into a set of hard and fast rules too develop a personal anointing and subsequently please God. Although God created physics - the various laws that govern the operations of this natural world, He is still a God predominantly focused on our relationship with Him. Too often we become involved in performing tasks for God when He would rather have us sitting at His feet.

As we conclude this effort to convert physical laws to spiritual principles, please keep these points in mind:

- **You cannot please God without faith.**
- **You cannot utilize faith that you have not developed.**
- **Faith not released results in anointing not performing.**

- **You cannot bowl over any symptoms or adverse circumstances without spiritual momentum.**

You do not need to comprehend all of the formulas in this book to understand these above statements.

Keep in mind that God produced this entire *ordered* universe from the Third Heaven dimension where He has established His throne. Therefore, we know that Heaven, at a minimum, is equally *ordered*.

If we will learn and apply the spiritual laws that function in the Kingdom of God, then we can manifest Heavenly results. In these end-times in which we currently abide, God is pouring out revelation of these spiritual laws as never before. Let's seek to discover and apply them.

ABOUT THE AUTHOR

Jack Shoup is the founder and senior pastor of Grace Fellowship of Georgetown located in Georgetown, Kentucky.

He is a graduate of The University of Tennessee in mechanical engineering and has an MBA from Xavier University. Beginning in 1977, Jack spent 14 years with IBM Corp. and another 2 years with Lexmark Corp. working in both engineering, business office, and management positions.

Jack and his wife Patty were both saved the same evening soon after they were wed in 1984. From that moment, Jack was imparted an insatiable hunger for the Word of God. He was called into full-time ministry in 1993 and has served as Pastor of Outreach Ministries, as an Assistant Pastor, and since 2001, as a senior pastor.

Jack is called of God to search out the Scriptures and to impart, through his teaching, in-depth understanding of the spiritual principles that govern the Kingdom of God. His primary area of instruction has focused much on the love of God and the covenant that believers enter into as members of the Body of Christ.

He has always had the drive to study the Word of God from an analytical standpoint believing that God is not unpredictable, but as a God who cannot lie, He is totally reliable. And, as a God who is reliable, His Kingdom must also be reliable, predictable, and the results of faith repeatable.

Jack and Patty have two daughters, Amy and Lori, both grown and pursuing their own careers.

OTHER BOOKS BY THE AUTHOR

The Supernatural Power of Love

Check-Mate

Eight Steps to Fulfill your God-Given Destiny

www.ingramcontent.com/pod-product-compliance
Lightning Source LLC
Chambersburg PA
CBHW050204130526
44591CB00034B/2095